The Military in
San Diego

THE MILITARY IN
SAN DIEGO

Scott McGaugh

Copyright © 2014 by Scott McGaugh
ISBN 978-1-4671-3156-8

Published by Arcadia Publishing
Charleston, South Carolina

Printed in the United States of America

Library of Congress Control Number: 2013951092

For all general information, please contact Arcadia Publishing:
Telephone 843-853-2070
Fax 843-853-0044
E-mail sales@arcadiapublishing.com
For customer service and orders:
Toll-Free 1-888-313-2665

Visit us on the Internet at www.arcadiapublishing.com

To those who serve America in uniform and all who support them.

CONTENTS

ACKNOWLEDGMENTS

A wide range of historians, military professionals, civic leaders, and volunteers dedicated to preserving the legacy of the military in San Diego made this book possible. Rudy Shappee provided vital baseline research at the outset. Bruce Linder, executive director of the Coronado Historical Association, and the USS Midway Museum's ship historian, Karl Zingheim, patiently proofed the manuscript. Lauren Rasmussen, Chris Travers, Carol Meyers, and Natalie Fiocre at the San Diego History Center provided indispensable access to a treasure trove of photography that forms the bulk of this book. Other photography donors included Ann Dakis and Tom LaPuzza of Space & Naval Warfare Systems, San Diego; Stephanie Venn-Watson and Mark Xitco of the US Navy Marine Mammal Program; and Andre Sobocinski of the US Navy's Bureau of Medicine and Surgery. Katrina Pescador, Debbie Seracini, and Alan Renga at the San Diego Air & Space Museum were especially accommodating, as was Kevin Sheehan of the Maritime Museum of San Diego. Elisa Leonelli, Faye Johnson, and Jan Kocian graciously provided photographs. Others, including Scott Sutherland and Sonja Hansen, were helpful as well. Unless otherwise noted, all images appear courtesy of the San Diego History Center.

INTRODUCTION

The history of the San Diego region is inextricably interwoven with the history of the US military. Over the course of five generations, this symbiotic relationship, which predated the birth of the nation, is one in which most San Diegans take a great deal of pride.

More than 200 years ago, the Spanish recognized the value of San Diego as a strategic military foothold on the West Coast. When California became a province of the United States in the mid-1800s, the nation quickly recognized San Diego's strategic position, with the area first being used as border protection. Then, as America turned its attention to the Far East, San Diego became even more relevant. City leaders worked with Congress and President Roosevelt in the mid-1900s, successfully advocating a "military metropolis" that eventually evolved into the San Diego region of today. Beginning with the arrival of the Great White Fleet in 1908, the military has been woven into the fabric of the region's fate.

As the armed services evolved in the 20th century, San Diego matured as well. The military presence spawned massive migrations of military personnel and civilians to the region, requiring vast expansions of housing, schools, public services, education, and health care. From welders to engineers, much of San Diego's labor force has been fueled by the employment needs of the military. Similarly, as technology evolved, San Diego's economic base has reflected shifts from aircraft manufacturing to military and defense space technology. San Diego's robust innovation sector, from high tech to biotech, as well as its world-class academic institutions, has been nurtured by billions of dollars in military research and defense (R&D) funding that has flowed into this region.

Today, military spending in the region is approximately $20 billion annually, the majority of which supports approximately 130,000 jobs. In addition, more than an estimated 300,000 civilians support the military's presence.

The relationship between the military and San Diego goes far deeper than the gross regional product. It speaks to the tens of thousands of Americans who depart from San Diego's shores every year in service to their country. It is reflected by military housing complexes throughout the region, where the children of civilians and those in uniform attend the same schools. Further, it is not just an active duty military community. San Diego is home to the third-largest population of veterans in the United States; 1 in 12 employees is a veteran, and 1 in 6 businesses is owned by a veteran.

Over the years, San Diego has evolved from a remote Spanish presidio outpost to a Pacific powerhouse and has become home to the largest Navy complex in the world. Worldwide commands of US Navy Special Forces, its surface fleet of ships, and all air wings are headquartered in San Diego.

Looking ahead, San Diego will continue to lead the nation in military/civilian stewardship and will continue to flourish as a community through its long-standing relationship with the armed forces of the United States.

—Charlotte Cagan
Executive Director
San Diego History Center

While San Diego may be internationally recognized today for its intimate relationship with the US Navy, the historical roots of its relationship with the military were first seeded by the US Army. Then nearly half a century later, the Army established the first substantial military aviation presence in the region. As national priorities and world affairs evolved, the Navy gradually became dominant. Here, marines conduct a marching demonstration for the public in 1915. Today, San Diego's relationship with the military spans not only the Navy, but also includes a vast array of military R&D institutions and alliances.

OPPOSITE: When the Spanish entered San Diego Bay in 1769, they built a presidio, or fort, on a hill overlooking the bay. The original fort featured two brass cannons and a wooden stockade surrounding a number of simple brush huts. The 300-by-300-foot compound later was reinforced with adobe blocks. After Mexico won its independence, the presidio became the residence of the governor of Alta California. The compound was abandoned in 1835 when its occupants moved into the new Pueblo de San Diego at the foot of the hill.

THE CHANGING GUARD
1769–1907

Over the course of the first 150 years of its history, San Diego's relationship with a local military presence underwent a number of changes. In 1769, the Spanish established the first military outpost to protect a nearby mission as well as to deter exploring sea captains from claiming territorial ownership for their kings and queens.

Spanish military influence continued until the conclusion of the Mexican-American War in 1848. Navy sailors captured San Diego at the beginning of the war, constructed a small fort of logs and earthen embankments, and named it Fort Stockton.

The subsequent Army-dominated period saw an increased military presence. Point Loma was recognized for its military value, and San Diego grew as a port of entry for supplies destined for the region's mostly Army outposts. In the latter half of the 1800s, the focus shifted to improving regional infrastructure to support a more diversified military presence. A river had to be relocated, and the bay had to be dredged.

By 1907, San Diego had matured from Spanish outpost of occupation, to a permanent Army foothold, to increasing Navy interest. This formed the backdrop for what would become the dominating local emergence of the US Navy.

Soon after the United States' declaration of war against Mexico in 1846, a force of American sailors and Marines aboard the USS *Cyane* rowed ashore and raised the American flag in the plaza of San Diego. After a series of skirmishes with Mexican loyalists, the ship's commanding officer, Comdr. Samuel DuPont, ordered his men to construct a fort above the old presidio. It marked the official possession of San Diego by the American military. The fort was later named Fort Stockton to honor Commodore Robert F. Stockton, the commander of the Pacific Squadron. A small garrison manned this earthen-walled fort until it was relieved by the arrival of the famed Mormon Battalion that had marched 2,000 miles from Iowa. The walls of the fort can still be seen near a monument to the Mormon Battalion atop what is now known as Presidio Hill.

After their arrival in San Diego in January 1847, members of the Mormon Battalion built a kiln to produce bricks for lining the town's wells. When off duty, the soldiers constructed the first fired-brick building in California. It was used at various times as a school and courthouse before burning down in 1872. This replica was constructed in 1992. (Photograph by Rudy Shappee)

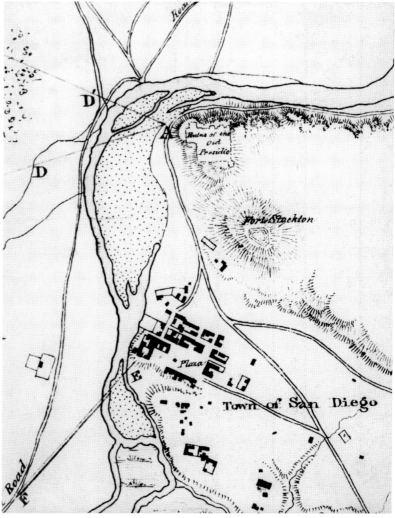

After the Mexican-American War ended, the military focus on San Diego waned as the US Boundary Commission laid out a new border between Mexico and the United States. This map of what later became known as "Old Town San Diego" shows the relationship of the locations of the old Presidio, Fort Stockton, and the town.

By 1874, San Diego had become a regional supply depot but was hampered by a shallow bay marked by mudflats. Most of the supplies had to be rowed ashore in small boats or ferried up the San Diego River, which still flowed into the bay at the time. Better waterfront access would have to be developed in the coming years for San Diego to prosper.

Land speculators such as William Heath Davis were eager to influence the military's plans for the area. He deeded two city blocks in his development called New Town to the federal government in 1850. Davis hoped that by having the Army construct its supply depot in his new development, he could sell additional land to various commercial interests.

The new depot consisted of a barrack block, featuring barracks, officers' quarters, offices, storehouses, and shops, as well as a corral block containing stables, hay barns, a watering trough, and a blacksmith shop. They are in the upper, right-hand corner of this photograph. A wharf extends into the bay, where a ship is moored in order to off-load its cargo.

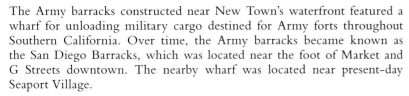

The Army barracks constructed near New Town's waterfront featured a wharf for unloading military cargo destined for Army forts throughout Southern California. Over time, the Army barracks became known as the San Diego Barracks, which was located near the foot of Market and G Streets downtown. The nearby wharf was located near present-day Seaport Village.

Life at the San Diego Barracks was not difficult for the soldiers stationed there. Some wrote letters home describing the mild climate and the pace of the work at the depot. A single company of soldiers off-loaded supplies from transport ships, loaded them onto wagons or pack animals, and then drove supply trains overland to Army installations at Fort Tejon, Fort Yuma, Fort Mojave, San Luis Rey, Chino, Santa Ysabel, and San Bernardino. By 1851, pack animals loaded with military supplies were a common sight, moving to and from the barracks area. But two years later, Davis's speculative land-development venture failed. Most business and residential buildings were moved to Old Town, a short distance away. By the end of the year, the Army barracks was almost the only major structure left in the short-lived New San Diego.

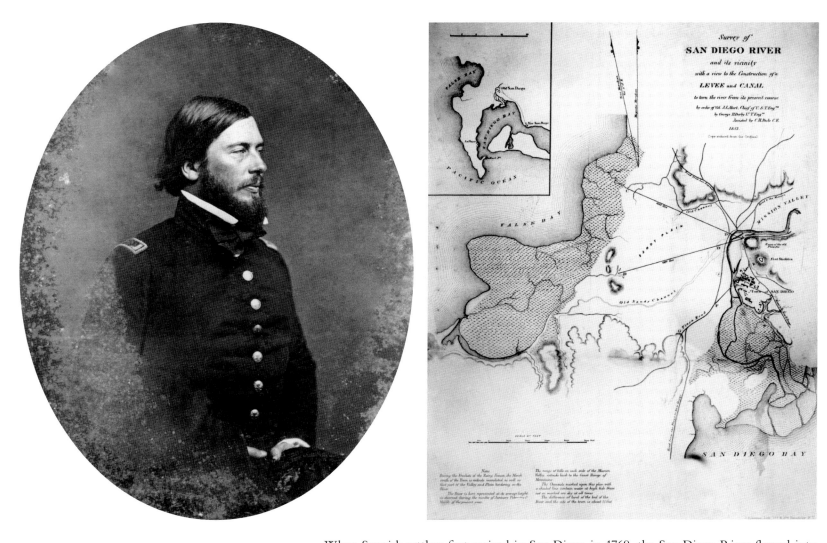

When Spanish settlers first arrived in San Diego in 1769, the San Diego River flowed into False Bay (now known as Mission Bay). Sometime during the 1820s, the river carved a new channel that emptied into San Diego Bay, filling vast portions with silt. The US Topographical Engineers sent Lt. George Horatio Derby (above, left) to San Diego in 1853 to develop a plan to divert the San Diego River back into False Bay. He designed two canals, each more than a mile long (straight lines on the map), to force the river's flow back into its original course westerly to False Bay and to stop the sedimentation of San Diego Bay.

Despite the tidal shoals caused by the outflow of the San Diego River, the center of San Diego Bay featured deep anchorages for ships arriving from the east. In 1867, Alonzo Horton started developing his "New San Diego" tract at the location of Davis's failed venture. Large wharves extended across the mudflats, separating the ships at anchor from the city's shoreline, which had an emerging commercial business district and military supply depot facilities. But it was a slow process. To increase the speed of transfer, rail lines were laid on the wharves so small yard engines could pull or push cars laden with goods from ship to shore.

David Choate was another prominent San Diego businessman who served as a founding member of the new chamber of commerce. In addition to being the author of the first publication by the chamber extolling the virtues of the bay, Choate played an active role in the Washington, DC, campaign to secure government investments in the San Diego Bay area.

George W. Marston owned the only major department store in San Diego. When the city's more prominent businessmen decided to form the Chamber of Commerce of San Diego County in 1870, Marston was one of its first members. The organization would soon begin a letter-writing campaign to Washington, DC, in an attempt to turn San Diego's harbor into a major commercial port.

Long wharves into San Diego Bay were not a permanent solution for commercial or military development. They were neither efficient nor high volume. The chamber of commerce issued bonds to pay for the dredging of some of the shallowest areas along the bay front. Pilings were sunk to create a bulkhead. Then the sediment was dredged and deposited along the shoreline. The net effect was to create a deeper anchorage closer to shore and to provide solid new ground for development. This became the start of decades of bay dredging, a significant portion of which was paid for by the federal government.

By the turn of the 20th century, San Diego's citizens had become used to the presence of destroyers and cruisers of the Navy's Pacific Squadron in the bay, including the crew of the USS *Bennington*. On July 21, 1905, a boiler aboard the armored cruiser exploded, killing 66 of its 111 officers and men. San Diego's citizens immediately aided the wounded sailors. When both local hospitals were filled to capacity, the old Army barracks was opened as an emergency treatment center. San Diegans donated blankets, pillows, bedding, food, and tobacco as the local hospitals' supplies dwindled. The dead were buried at the Army cemetery atop Point Loma. Today, a tall monument marks the site of their graves.

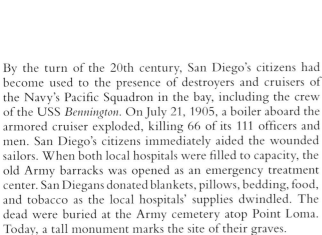

The federal government designated Point Loma as a military reservation in 1852. Construction of harbor-defense gun emplacements began in 1853 but was halted due to lack of funds. The soldiers were initially housed in this tent encampment on the bay side of the peninsula while awaiting the construction of suitable barracks. Permanent development of modern harbor defenses began in earnest in the 1890s among mounting concerns of potential war with Spain. This included rifled cannons mounted on disappearing carriages at the mouth of the San Diego Harbor. In 1897, construction began on an emplacement of four 10-inch breech-loading rifles and two separate emplacements of two 3-inch rapid-fire guns, each on Ballast Point. A mining casement, storage facilities, and a wharf for loading mines were also authorized. By 1898, Company D, 3rd US Artillery had arrived to occupy the new fortifications.

Early-day Point Loma artillerymen practiced with Civil War–era unrifled Napoleon cannons. Artillery fire rattled the windows of the homes and businesses across the bay as the artillerymen practiced firing at targets mounted on barges in the ocean outside the bay. They also were used to fire special salutes to passing ships and during morning and evening colors ceremonies. Over time, powerful weapons hidden within thick reinforced concrete casements such as these were installed on Point Loma. The larger and better-protected weapons added both range and firepower to the defense of San Diego Bay. For a time, mines were laid across the mouth of the harbor as well.

Permanent structures were constructed to house the officers and men of the Army post. By 1903, enough buildings had been constructed to justify the transfer of the garrison from the San Diego Barracks to Point Loma. The garrison consisted of 7 officers and 211 enlisted men. A small fort with two 3-inch guns also sat opposite the Point Loma batteries on North Island. (National Archives.)

The new Army installation was named Fort Rosecrans after Gen. William S. Rosecrans. He had visited San Diego shortly after the war and briefly speculated in real estate, owning the city block bounded by F and G Streets and Fifth and Sixth Avenues before selling it to developer Alonzo Horton for $2,000. (Library of Congress.)

In the early 1900s, Fort Rosecrans troops periodically assembled for field inspections when the Department of California's commanding officer, Brig. Gen. Tasker Howard Bliss, visited. Bliss also inspected the outpost's rapid expansion of cannon firepower. The various cannon batteries were named after noted artillerymen, including some who had served in the Civil War and another who had earned the Medal of Honor. Fort Rosecrans offered some of the most desirable duties in the Army in the first decade of the 1900s. Some called it "the old soldiers home," given the number of experienced soldiers who secured transfers to Fort Rosecrans. Many retired from the Army after serving at the outpost and remained in San Diego, where they became prominent citizens.

For years, the chamber of commerce lobbied for a greater military presence in San Diego as a method of economic development. An early chamber initiative convinced the Navy to construct a ship refueling station next to the quarantine station on Point Loma in 1904. That marked the start of a subtle shift from Army dominance to a greater Navy presence in the area. William Kettner arrived in San Diego in 1907 to establish an insurance business. He quickly became the latest in a line of visionary San Diego entrepreneurs who took an active role in the chamber of commerce. Within a year of his arrival, he was elected to head the chamber of commerce. That placed him in a position to orchestrate future campaigns for an even more pervasive military presence in the region.

Point Loma was ideally suited for the Navy's need for a high-powered radio station to communicate with its expanding Pacific Squadron. In 1905, the Navy was authorized to construct a state-of-the-art "wireless radio telegraphy station" atop Point Loma. On May 12, 1906, a chief petty officer and two sailors delivered this Massie 5-kw transmitter-receiver to the new station.

The transmitter-receiver represented the new age of long-range wireless communications between ships and land-based stations around the world. Once operational, the Navy developed the site with these additional antennas, adding range to the expanded facility. These tall structures became a signature image of Point Loma for decades to come.

One of the unsung drivers of expanded military presence in San Diego was H.P. Wood (at left). During his term as chamber of commerce secretary (1899–1905), he was sent to Washington, DC, to successfully lobby for a Navy refueling station at the mouth of the bay. Around the same time, he played a key role in a fundraising campaign to establish what became known as the Scripps Institution of Oceanography. The emerging triumvirate—including the San Diego business community, Scripps, and the military—would go on to dramatically influence the development of San Diego over the next 100 years.

OPPOSITE: The Great White Fleet armada of 16 frontline battleships was scheduled to largely bypass Southern California during its 1908 world cruise. Only in Los Angeles would the fleet's four divisions separate and briefly stop at four locations. San Diego civic leaders had grander visions and lobbied the secretary of the Navy for a port of call by the entire fleet. Frustrated at his lack of response, they hired a boat and sailed south to Magdalena Bay, where they convinced Adm. Robley Evans to make a four-day stopover in San Diego.

TWO

AN ERA OF OPPORTUNITY
1908–1916

San Diego remained a largely isolated city at the turn of the 20th century. Campaigns by city leaders dating back to the 1870s to diversify its economy and generate city growth had yielded little. Then, a confluence of three nearly simultaneous developments produced an era of opportunity. Each could greatly influence the city's future by potentially broadening its economic base, enlarging San Diego Bay, attracting newcomers, and generating national attention.

First, America was turning its attention to the Far East. When Pres. Theodore Roosevelt sent his Great White Fleet in 1908 on a world cruise to demonstrate American power and purpose, its planned route up the California coast caught the city's attention. Secondly, the nation, including the military, became captivated by the potential new technology called the airplane. Finally, as completion of the Panama Canal neared, its future impact on commercial accessibility to California and the Pacific was difficult to underestimate.

Could San Diego profit from these developments? Would aviation pioneers see the advantages of San Diego's climate and accessible open space, both of which were perfect for a year-round airfield? Could the Great White Fleet make a port of call in San Diego? How could the chamber of commerce and others make the Panama Canal benefit San Diego? The answers to these questions would define the San Diego of the 20th century.

On April 14, 1908, division flagship USS *Georgia* arrived at its Coronado anchorage. All hulls in the fleet were painted white to reflect the sun's heat, making them a dramatic sight as they approached. The ships were met by small boats crewed by San Diegans that were filled with flowers and lemonade for the fleet's 14,000 sailors.

The Navy's newest 14,000-ton battleships, with a draught of 24 feet, could not enter the shallow San Diego Bay. After the ships dropped anchor off Coronado, their crews faced a 10- to 12-mile trip in small liberty boats into the bay and to the long wharves along the embarcadero. (Maritime Museum of San Diego.)

Many visiting sailors came ashore near the foot of Broadway. Once ashore, the crews were treated to free movies, food, and transportation on the city's trolley cars to almost any city destination they chose. Recreational shooting galleries were especially popular among the enlisted men. Most wore souvenir buttons or streamers presented by local residents. Meanwhile, their officers were entertained at teas, sumptuous dinners, and dances at the Hotel del Coronado and the US Grant Hotel. The San Diego business community wanted to demonstrate San Diego's hospitality and eagerness to host a permanent military presence. City leaders had raised $20,000 from San Diego's 40,000 residents to finance the four-day reception. Admiral Evans was too ill to attend any of the San Diego festivities and was replaced as the Great White Fleet's commanding officer when it reached San Francisco.

San Diego's Great White Fleet parade of more than 5,000 sailors and officers was called the largest naval parade on soil since the Civil War. Organized by the fleet celebration committee on relatively short notice, it was a two-mile march from the foot of Broadway, up a gentle rise under sunny skies, to Balboa Park. More than $6,000 in bunting adorned office buildings along the route, and estimates of the crowd ranged from 50,000 to 75,000, nearly twice the city's permanent population. Women's organizations served BBQ and lemonade at the conclusion of the parade. (Both, Maritime Museum of San Diego.)

Night Illuminations of Battleships off Coronado

The relatively new invention of electricity was put to good use at many of the Great White Fleet's ports of call, including San Diego. In San Diego's case, the fleet stood at anchor in a line a short distance offshore. Most nights, the superstructures and masts of the fleet were dramatically illuminated. Nightly searchlight displays sweeping the sky added to the drama. Similarly, newly installed city streetlights and individual business lights created a dramatic, never-before-seen effect. The officers presumably slept well on their brightly lit ships after days spent on tours led by city leaders intent on promoting prospective Navy stations, dry docks, and training areas. One of those leaders, William Kettner, later became the area's congressman in 1912 and continued to press for an expanded and permanent Navy presence in San Diego.

The armored cruiser USS *California* was renamed the USS *San Diego* in 1914. It became the only major American warship lost in World War I, likely from a German mine or torpedo off the coast of New York. This was the first of four ships to be named the USS *San Diego*, the fourth as recently as in 2010. (Maritime Museum of San Diego.)

Crews were eager to demonstrate the capability of the fleet's 8-inch and 12-inch guns. But the reality was that this massive firepower made the Virginia-class battleships top-heavy and produced an uncomfortable tendency to roll. Ultimately, the 8-inch guns proved unusable due to their proximity to the 12-inch guns.

In the years following the Great White Fleet's arrival, San Diego Bay's dredging continued. Overly long piers built in the 1800s across tidal mudflats were not adequate for military operations. An expanded military presence in San Diego depended on greatly improved bay access and far more permanent piers. In 1913, this dredging focused on creating Harbor Drive along the downtown embarcadero. That would become the home for the Broadway Pier, Navy Pier, and a major Navy supply depot operation on Harbor Drive. For decades, trains would roll out onto Navy Pier and transfer ammunition and supplies onto Navy ships. But none of that would have been possible without the years of dredging during San Diego's "Era of Opportunity."

Lt. Theodore "Spuds" Ellyson became the Navy's first aviator when he graduated from the Curtiss School of Aviation. The former submariner in 1912 opened a Navy flying school nearby called Camp Trouble. Within six months, all of his school's aircraft had been destroyed in crashes. The Navy abandoned North Island and naval aviation did not return to San Diego until 1917. (Glenn H. Curtiss Museum.)

When he turned his attention to aviation, Glenn Curtiss was already famous as a motorcycle racer. In 1910, he sought a year-round seaplane operations facility for his New York aircraft manufacturing plant. In early 1911, he opened an aviation school on North Island, charging $500–$600 per student. Tuition could be applied to the purchase of one of his aircraft. (Library of Congress.)

Curtiss Aviation Field
North Island, San Diego 1911

When Glenn Curtiss arrived in San Diego in 1910, he sought "a place not easy of access to the curious crowds that gather wherever there is anything novel to be attempted; for a flying machine never loses its attraction to the curious." He found it on North Island, where he leased an expanse of sagebrush from the Coronado Beach Company. North Island's bay side shoreline was perfect for developing the "hydroaeroplane." He offered free training to Army and Navy pilots as an incentive to attract military interest in his aircraft. The Army sent three officers to Curtiss and then established an aviation school on North Island that would last for 26 years, even though 12 of its first 48 pilots were killed in flying accidents. The string of controversial fatalities led to a *San Diego Union* headline that read, "Intrepid Navigators of Air Crushed, Mangled to Death in Fall of Government Biplane."

Curtiss experimented with various amphibious aircraft designs when classes were not being held. Numerous test flights were conducted over San Diego Bay in early 1911, to the delight of thousands of San Diego onlookers. Initial designs with two pontoons were soon replaced by single-pontoon aircraft. The pontoon was 12 feet long, 2 feet wide, and 1 foot deep. Supporting slats on each wingtip kept the wings out of the water. Evaluations were made of aircraft with engines behind the pilot and those with motors in front (causing numerous exhaust and sight challenges for the pilot). Some students taking the more expensive hydroaeroplane course enjoyed chasing pelicans across the bay. Here, Lt. John C. Walker Jr. lies on the pontoon as assistants prepare a hydroaeroplane for takeoff with Curtiss at the controls. Walker was one of first three Army aviators taught by Curtiss, selected from 30 Army volunteers.

Curtiss established a navigation milestone when one of his aircraft successfully landed on a makeshift landing deck aboard the USS *Pennsylvania* in San Francisco Bay in January 1911. But temporary modifications interfered with the armored cruiser's guns. A few weeks later on February 17, Curtiss convinced the USS *Pennsylvania*'s captain to participate in a demonstration of his hydroaeroplane in San Diego Bay. Curtiss landed alongside, stood on the top wing when the aircraft was hoisted aboard, and then his aircraft was lowered back onto the water for the flight back to North Island. The USS *Pennsylvania*'s firepower remained unaffected. At that moment, naval aviation was born. Three months later, the Navy bought its first aircraft from Curtiss. Congress later designated the city as the birthplace of naval aviation.

As the Mexican revolution intensified in early 1911, uncertainty along the US border mounted. Detachments of Fort Rosecrans soldiers used the San Diego & Arizona Railroad to reach remote encampments. They were part of the 20,000 troops Pres. William Howard Taft ordered to patrol the border. On June 22, 1911, a total of 105 *insurrectos* were interned at Fort Rosecrans. They had been driven across the border by Mexican federal troops who had regained control of Tijuana. Of the 105, twelve were found to be American military deserters, and ultimately, three were turned over to a US Marshall. By June 30, the provisional brigade that had been formed to patrol the border was disbanded and most troops had returned to their regular duty stations.

In 1912, San Diego and San Francisco were locked in a battle to host the Panama-California Exposition to celebrate the upcoming opening of the Panama Canal. Pres. William Howard Taft openly supported San Francisco's bid by inviting other countries to only participate in that city's exposition. But national elections that year turned the tide. San Diego civic leader William Kettner was elected to Congress, and Woodrow Wilson became president. Although Wilson lost the California vote, he carried San Diego by 1,600 votes and counted Kettner as a key local ally. In gratitude for Kettner's support, on May 23, 1913, Wilson signed Kettner's legislation that authorized federal agencies to permit free admission to their exhibits at the San Diego exposition. Less than six weeks later, San Diego voters, by a margin of 16 to 1, authorized $850,000 worth of bonds to construct Balboa Park buildings for the exposition. They would showcase San Diego, including its strategic military potential, to the nation. (Library of Congress.)

The Panama–California Exposition opened on January 1, 1915, in Balboa Park. The organizing committee had been working on the exposition for five years and had raised $7 million. The night before, a kickoff celebration including eight massive searchlights on the USS *San Diego* (moored off Market Street) illuminated Balboa Park's tower, more than a mile away. One of the first sights thousands of expo visitors crossing Puente Cabrillo (now commonly called the Laurel Street or Cabrillo Bridge) witnessed was a "model Marine Camp." It included "a well nigh perfect parade ground, athletic field and city of neat tents." More than 100,000 people attended the exposition in the first week. By the end of 1915, more than two million (in a city of 50,000) had witnessed a variety of military encampments, demonstrations, and exhibits. The Navy set several recruiting records at the exposition as well.

In late April 1915, the Marines conducted a "sham battle," firing blanks for visitors, including renowned filmmaker Sigmund Lubin, who was interested in establishing a studio in San Diego. The demonstration was part of a prominent role played by the Marines, Navy, and Army during the yearlong celebration. Dignitaries regularly were treated to cavalry demonstrations of horsemanship, military parades, and concerts by military bands. Several dignitaries clearly were impressed. On March 28, 1915, Assistant Secretary of the Navy Franklin Delano Roosevelt arrived in San Diego and announced that the Atlantic Fleet would be coming to San Diego and promised that the city would become a supply and liberty port for the Navy. Roosevelt also gave Congressman William Kettner the impression he favored the climate of San Diego over San Francisco for a future sailor training station.

On July 27, 1915, former president Theodore Roosevelt was one of many national leaders who attended the exposition. He advocated for a 200,000-man standing Army in a speech at the Organ Pavilion. Two days earlier, he had stated that "an antimilitary nation that is powerless to help itself is powerless to further the cause of humanity." (Library of Congress.)

Marine Corps colonel Joseph H. Pendleton became an enthusiastic San Diego ambassador during the exposition. He believed "San Diego's every advantage of climate, of strategic location, of wonderful natural formation of land and sea, make it the perfect, the ideal location for a Marine Corps Station." He envisioned North Island as a suitable site, not the remote coastal tract to the north now called Camp Pendleton.

The Navy exhibit in the Commerce and Industries Building, not far from Balboa Park's renowned lily ponds, was especially popular. The exhibit included torpedoes, diving suits, machine guns, and models of the USS *San Diego* and USS *North Dakota*. It marked the start of an ongoing Navy presence in Balboa Park. At the end of the exposition, Congressman Kettner offered the Navy several exposition buildings as training facilities for $1 per year. They would become the Naval Training Station San Diego and the basis for the first Navy medical facility in San Diego. The exposition became a springboard for a much wider Navy presence in San Diego. Fundraising campaigns were launched to buy land in San Diego for Navy use. Four years later, construction on a permanent Navy hospital began, nearly contiguous to the exposition site. These lily ponds would be used for a variety of military purposes in World War I and World War II. (National Archives.)

Visitors experienced a taste of military life both at the exposition and throughout San Diego. Daily marching demonstrations, such as this in the Plaza de Panama, added a touch of pomp and circumstance. Antiaircraft artillery fired blanks at aircraft from North Island's Rockwell Field. Navy ships conducted target practice at barges towed a few miles off the coast. The USS *Colorado* gun crew won a competition and was awarded $2,000. Artillery demonstrations at Fort Rosecrans thundered across the city. San Diegans were impressed. San Diego mayor Edwin Capps later wrote, "I feel that I am expressing the sentiment of the people of our city when I say that we very appreciatively acknowledge our obligation to (the military) for contributing to the success of our fair and for the cordial co-operation which has always been given . . . by the men of the Army and Navy." City voters approved several transfers of land to the Navy for the development of naval facilities, in one case by a 13,857-305 margin.

Ellen Browning Scripps played a key role in forming what became the Scripps Institution of Oceanography and national defense partnership of today. After moving to La Jolla in 1896, the wealthy newspaper publisher pledged $50,000 to establish a permanent biological research lab comparable to the Strazione Zoologica in Naples, Italy. She also financed her family's purchase of a large ranch a few miles east of La Jolla.

Edward Willis Scripps was not quite as enthusiastic as Ellen, but he also pledged seed money for the institution that now bears his family's name. His family later sold the nearby 2,130-acre ranch that he called Miramar to the federal government in World War I. It became an infantry training site and later the original Navy Fighter Weapons School, better known as Top Gun.

The first marine biological lab in San Diego was temporarily housed in the Hotel del Coronado's boathouse. The Scripps family and a few local biologists had a more ambitious vision. They envisioned a permanent research station north of San Diego, near Long Beach (known today as La Jolla Shores). Scripps family money was used to purchase 170 acres from the City of San Diego on an isolated bluff overlooking the sea. Only the structures shown here existed in 1912. Shortly after this photograph was taken, 12 cottages, an executive director's house, a library, a 1,000-foot-long pier, and an 85-foot ship, the *Alexander Agassiz*, were constructed. "I have never . . . asked either Mr. Scripps or Miss Scripps for money. All I have ever done has been to point out the possibility of development and research and the meaning of the results," wrote William Ritter, one of the first local biologists to promote a San Diego marine research station.

OPPOSITE: The Marines' 4th Regiment remained encamped in Balboa Park following the Panama-California Exposition. As Col. Joseph Pendleton lobbied for a permanent San Diego Marine base, the Marines' former location on North Island was deemed too expensive. A survey of potential locations identified a 232-acre parcel called Dutch Flats at the north end of San Diego Bay as a viable permanent location. But until federal funds were authorized and construction was completed, the Marines remained in Balboa Park.

BIRTH OF A
MILITARY METROPOLIS
1917–1921

For decades, San Diego leaders had staked the city's future on a greater military presence. At every opportunity, they mobilized to buy the land coveted by the military and make it available, often at no cost. Yet prior to World War I, the military had not become the economic engine that drove San Diego.

On the eve of World War I, the city's voters again endorsed public policy supporting development. In 1917, they elected pro-growth Louis Wilde as mayor over George Marston,

who was seen by many as anti-business, by a 58-42 percent margin. World War I became an opportunity to forever shift San Diego's long-sought industrial development goals toward a military-industrial economy.

Over the next four years, the Navy sought to develop a greater West Coast presence. That presented a new opportunity for Congressman William Kettner and his San Diego business colleagues. They saw the Navy's desire as a springboard for local economic prosperity. Their

strategy was straightforward: make San Diego too attractive for the Navy to ignore.

The resulting series of land grants to the Navy during this period gave rise to several military institutions that remain today. The birth of the San Diego military metropolis in this period led to a world-renowned military hospital; a major Marine Corps recruiting and training presence, the nerve center of naval aviation; and a massive defense science and technology economy.

14th Co'd Dining Room.

Although the US Navy took over most of the buildings in Balboa Park following the Panama-California Exposition, tent encampments remained necessary when the Navy presence grew substantially throughout San Diego during World War I. Sailors patronized downtown restaurants and various entertainment establishments. Dozens of warships dropped anchor in San Diego Bay while the skies sometimes were filled with Army and Navy aviators who had taken off from multiple military aviation training facilities on North Island. More permanent Navy facilities were needed. Following the war, the Navy was deeded more than 250 acres to establish a naval training station at Loma Portal on San Diego Bay. The Navy also received more than 18 acres near Balboa Park for a $2.4 million naval hospital. As a measure of local support, San Diego voters ratified land donations for major naval developments by large margins. For instance, the vote for hospital land passed 9,289-137.

Sometimes open to the public, the Plaza de Panama was where various sailor drills, presentations, and ceremonies were conducted in World War I by the 4,000 Navy recruits who came from the western states. Due to wartime conditions, when their training was complete, most sailors skipped ship-orientation cruises and were immediately sent to their assigned warship.

Balboa Park's buildings were uniquely sized to accommodate an expansive Navy operation that amounted to a self-contained city. The former Cristobal Café fed up to 5,000 sailors in a single seating. The *San Diego Union* speculated that Balboa Park could become the Navy's permanent base if a nearby canyon was dammed to create a 1,200-by-200-foot lake for sailor training.

Thousands of sailors slept in the Commerce and Industry Building. Recruits also used several former exposition buildings for classroom instruction, small arms training, physical fitness drills, meals, haircuts, purchases at the canteen, specialty training, and lectures such as "Life and Duties of the Man-o-War's Man." As early as 1916, Congressman William Kettner proposed the establishment of a permanent Navy training station in San Diego. In 1919, the chamber of commerce raised $280,000 to buy bay side land at Loma Portal (next to the Marines' planned base) for transfer to the Navy as a training station site. In addition, the city granted 142 acres of tideland. Congressman Kettner later secured a $1 million federal appropriation for construction of the naval training station.

Thousands of sailor recruits, 40 percent of whom could not swim, were trained in Balboa Park, three miles from the ocean. San Diego did not have a large municipal pool. The solution was to take over the park's renowned lily pond in 1918. It was deepened, lined with concrete, and a new water system exchanged its 450,000 gallons daily. The pond was too small to learn small-boat handling skills, so 25-foot cutters were tied stationary, enabling the sailors to practice rowing in the stationary vessels. In World War II, the pond was used as a rehabilitation facility by the Navy's nearby hospital. (Both, San Diego Library.)

On May 30, 1917, Secretary of the Navy Josephus Daniels announced Navy plans to build an air station at North Island. Rockwell Field opened in September when the Navy returned to North Island, alongside the Army Signal Corps Aviation School. The Navy took residence on the bay side of North Island to allow access to piers while the Army moved to the ocean side. By that time, President Wilson had signed an executive order authorizing condemnation of North Island for military use. But it was not until 1919 that Congress appropriated $6 million for acquisition. Meanwhile, the Army taught Signal Corps cameramen how to use cameras mounted on the outside of their fuselage while others built JN-4 aircraft in newly constructed hangars during World War I. Learning to fly at that time was a dangerous and expensive business. Engines were commonly salvaged from wrecked aircraft and reused. (Library of Congress.)

ALL AMERICAN
LA PERE (ABOVE)
ROCKWELL FIELD SAN DIEGO

CAPTURED
GERMAN FOKKERS

By the time of the armistice, nearly 2,000 Army and Navy pilots and mechanics had been trained at North Island. At the end of the war, the Navy air station included nearly 500 officers and enlisted men and nearly 500 aircraft. Following World War I, North Island received some of the more than 125 German Fokker aircraft (shown here) that had been captured in the war. They were used in a variety of postwar experiments that resulted in new aircraft designs, leading to several aviation milestones. In one instance, the first transcontinental flight (in a Fokker T-2) ended at North Island after 26 hours, 50 minutes, and 38 seconds' flying time. Today, Naval Air Station North Island remains a strategic part of the Navy's air forces worldwide.

The facilities for Camp Kearny troops produced a construction boom. Ultimately, 1,200 buildings were constructed, including a 1,000-bed hospital. By the end of the war, approximately 65,000 men had been processed through Camp Kearny. After the armistice, it was used as a demobilization center. Today, the Marine Corps operates an air station where Camp Kearny once stood.

America needed new massive Army training camps in the run up to World War I. San Diego officials saw that need as another opportunity for economic development. They offered the military an undeveloped ranch called Miramar, about 11 miles north of downtown, for an Army infantry training center. Soon, 40,000 troops were training at Camp Kearny and preparing for deployment to Europe.

"America's Sweetheart" Mary Pickford visited Camp Kearny on several occasions, usually to entertain the troops and entice them to buy war bonds. She was a part of a parade of Hollywood celebrities who visited Army training camps. Camp Kearny had a 3,000-seat multiuse building for large events, a 1,100-seat theater, and a 2,000-seat auditorium. Among those in the audience was Army draftee Buster Keaton, the popular Hollywood comedic actor.

Although Camp Kearny was north of the city, downtown parades were a popular and frequent sight. Rows of soldiers, 20 men wide, marched up Broadway, which was lined with civilians as many as 20 people deep on some corners. Following World War I, Camp Kearny was abandoned. Its airstrip, however, was used to test Charles Lindbergh's *Spirit of St. Louis* in 1927.

San Diegans had two methods through which they could help finance World War I. Just as President Lincoln had done in the Civil War, President Wilson called for a national wartime fundraising campaign. He sought $170.5 million in donations to designated charitable organizations such as the Salvation Army, National Catholic War Council, Jewish Welfare Board, and American Library Association. In addition, San Diegans were avid purchasers of Liberty Loan bonds, seen here in front of the Liberty Theater on F Street. The *San Diego Union* called bond purchases a barometer of the city's patriotism. Over the course of four bond campaigns, local bond sales amounted to $150 for each of the city's 100,000 residents. At the same time, San Diego benefitted from increased federal spending. The military committed approximately $19 million to the San Diego region during World War I.

The Marines moved into their partially completed base in 1921, not far from where recruits would begin arriving at the Navy's new training station the following year. A $1.5 million cost overrun on the Marines' $4 million construction project delayed completion. Only 60 percent of the planned project was completed when the first phase was finally halted in 1926.

World War I interrupted the Marine Corps' plan to build a permanent base on San Diego Bay following the Panama-California Exposition. In 1916, voters had approved a 500-acre donation to the Marines by a 40,288-305 margin. Construction did not start until 1919, when dredging for the site was 80 percent complete. Balboa Park's architect, Bertram Goodhue, also designed the Marines' base. (National Archives.)

The Navy sought to rebalance its postwar fleet and favored a single base on the West Coast. San Diego's too-shallow bay could not accommodate battleships. But Congressman Kettner and local leaders developed a plan to secure at least a portion the Pacific Fleet. Kettner terminated the lease of a shipbuilding company he had helped sponsor at Thirty-Second Street along San Diego Bay. Then the 98 acres of land, water, and marsh were given to the Navy, complete with a ship repair facility and rail access. With the support of Rear Adm. Roger Welles, the commander of the "naval district" that the Navy had established for San Diego, this land became the primary base for West Coast destroyers. By the early 1920s, a southern portion of San Diego Bay became the major Navy base that remains today. (US Navy.)

On June 28, 1910, the Navy's first two submarines, the USS *Grampus* and USS *Pike*, arrived in San Diego Bay, concluding the longest submarine sea voyage to date at the time (550 miles over 11 days). Both submarines conducted perilous drills, tested their capabilities, and sometimes were hauled ashore for repairs. Owing to the danger, the crew was paid an extra $5 a day, plus an extra $1 per dive. Practice torpedoes fired in the bay were retrieved by divers. On May 30, 1917, Secretary of the Navy Daniels recommended a submarine base be established in San Diego Bay. Although Navy submarines and their tenders put in at San Diego for several decades, a permanent submarine base was not established until 1959.

One of the earliest and most nondescript military installations in San Diego was Defense Fuel Support Point Loma, originally called the US Naval Coal Station. In 1900, city officials recognized a permanent fueling station with deep-water access near the entrance to San Diego Bay was critical to establishing Navy bases here. Although a pier-side facility technically was operational by 1904, it took approximately eight years before the coal station was fully functional. Then, as ships converted to liquid fuel and the Navy expanded its bases here, more than 50 aboveground and belowground tanks were installed on the 200-acre site, many of them in 1932 and 1939. In 2009, the Navy began a five-year overhaul of the site, where the storage capacity had increased to one million gallons of ship and jet fuel. (National Archives.)

The Navy often turned to North Island to pioneer or validate advances in aviation. In 1919, this hangar was constructed to house early Navy blimps and balloons, including the USS *Shenandoah* (ZR-1) when it visited San Diego. The Navy conducted blimp experiments to test their capability as spotters and for antisubmarine warfare. The final descent to a mooring tower was a perilous process for crews limited to hand signals. San Diego's prevailing breezes off the ocean added to the danger. A few years after the photograph of the Shenandoah docking at sea was taken, it crashed in severe weather in the Midwest with the loss of all hands. (Both, US Navy.)

In 1912, the Navy decided to construct a worldwide chain of radio transmitters. One of the first to be built was the Chollas Heights station, about five miles southeast of downtown San Diego. On July 21, 1914, the Navy bought the 75-acre site for its 200,000-watt communications system. The three 600-foot towers were commissioned in January 1917. The first message, tapped out in Morse Code on a silver key made by local jeweler Joseph Jessop, was received by Secretary of the Navy Daniels. The Chollas Heights Naval Radio Transmitting Facility became the most powerful transmitter in the world. (Library of Congress.)

OPPOSITE: In the two decades following World War I, San Diego evolved from a fledgling constellation of Navy bases to the epicenter of the Navy's war-fighting capability in the air. The arrival of the USS *Langley*, the Navy's first aircraft carrier, drove rapid expansion of the Navy's North Island infrastructure as pilots learned to launch and land on the "flattops." Here, a Vought UO-1 launches off the flight deck of a carrier, likely the USS *Langley*, while berthed pier-side at North Island. Pioneering naval aviation milestones came quickly.

FOUR

NAVY TOWN
1922–1939

San Diego fully committed its future economic growth development to the military between World War I and II. It came at a time when rising worries over Japan spurred a relocation of Navy resources to the West Coast, especially destroyers and accompanying cruisers—the kinds of ships San Diego Bay could most easily accommodate. Then came the explosion of military aviation technology. Over the course of nearly 20 years, military aviation evolved from primitive biplanes to sleek fighters, long-range bombers, and advanced seaplanes.

San Diego was positioned to take advantage of it all. Civic leaders stood ready to give away land and be the perfect hosts as the aircraft carrier era unfolded, the Navy and Marines sought new training facilities, and the West Coast took on greater strategic importance.

This, coupled with the advent of a military aircraft manufacturing base, made San Diego a true Navy town. As a result, steady military spending throughout San Diego buffered the local economy from the worst effects of the Depression. As World War II approached, San Diego had gone all in as a city devoted to the military.

Recruits at the naval training station were eager for liberty after spending their first three weeks quarantined in isolated pockets of tents to prevent the spread of chicken pox, measles, polio, and influenza. When off duty, many headed for downtown on city trolleys. Businesses on Broadway and the nearby Stingaree District welcomed them with arcades, theaters of various ilks, bowling alleys, and bars. Meanwhile, the chamber organized more formal affairs for the growing Navy presence. In 1937, the chamber's Army and Navy committee hosted 47 social events, usually dances, for the more than 20,000 sailors.

The skies over San Diego became a proving ground for daring aviators eager to test the limits of rapidly evolving military aircraft. On June 27, 1923, the first successful air-to-air refueling took place, about 500 feet above Rockwell Field. Army aircrews transferred 75 gallons of gas through a hand-held, 50-foot hose during a flight of more than six hours. Two months later, a world endurance record was set when 14 refuelings took place during a single flight, enabling a North Island crew to fly 3,293 miles over the course of more than 37 hours. As the capabilities of various military aircraft variants increased, so did aviation's strategic role in both the Army and Navy in the 1920s and 1930s. (San Diego Air & Space Museum.)

Ceremonial flyovers at various civic events became a common sight in San Diego. But Rockwell Field also became routinely congested, and clouds of dust plagued ground operations. In 1928, the Army designated Rockwell Field as its headquarters for the 7th Bombardment Group, 11th Bombardment Squadron, and 95th Pursuit Squadron. The Army also announced plans to construct new quarters, barracks, hangars, shops, and support facilities. Meanwhile, the arrival of aircraft carriers to San Diego brought additional squadrons of Navy aircraft to North Island. The Navy needed to expand, too. In 1929, an Army/Navy board was established to investigate the overcrowding. It recommended that Army air operations be transferred from Rockwell Field to another location. No immediate action was taken, however.

Throughout the 1920s, the Navy presence in downtown San Diego grew dramatically. By 1924, Navy ships were tying up to Municipal Pier at the foot of Broadway (now called Broadway Pier). In 1926, city planner John Nolen proposed the creation of Harbor Drive several hundred yards out into the harbor, a project that would require continued bay dredging. The Navy saw the new thoroughfare as an opportunity to efficiently move men and armament from its Thirty-Second Street base around the east side of the bay to Point Loma. A few years later, the Navy built a seven-story regional supply depot headquarters one block south of Municipal Pier as well as the 1,000-foot-long Navy Pier to service a variety of ships. This photograph from 1928 reflects the ongoing transformation of the downtown embarcadero to accommodate the Navy. Today, virtually all of the land between Pacific Highway and Harbor Drive is the result of bay dredging. (Maritime Museum of San Diego.)

INMAN CO.
A2712-F

SEMAPHORE DRILL
U.S. NAVAL TRAINING STATION
SAN DIEGO, CALIFORNIA.

On June 15, 1923, a total of 350 recruits and 65 instructors arrived at the new naval training station. The 16-week training regimen included semaphore (signaling with flags) training and then was expanded to include preliminary radio, yeoman duties, and, for some, bugler and band instruction. By 1929, the training station's schools included Cooks and Bakers, Sound School, Buglemaster, Musician, Radio Operator, Gyro Compass, and Stenography. Daily inspections taught recruits how to be "squared away" when they arrived on their first ship. Training was shortened to 7 weeks during World War II and then expanded to 11 weeks during the Korean and Vietnam wars.

From the beginning, Hollywood turned to San Diego as its de facto military back lot. As a result, more than 100 military movies have been filmed in San Diego. *Tell it to the Marines*, starring Lon Chaney in 1926, was the first movie made with the full cooperation of the Marine Corps. Scenes were shot at the Marine Corps Recruit Depot (MCRD). MGM brought in MCRD's commanding officer, Gen. Smedley D. Butler, photographed here inspecting authentic troops the following year, as the movie's technical advisor. Butler twice was awarded the Medal of Honor before retiring and became a vocal critic of American military intervention.

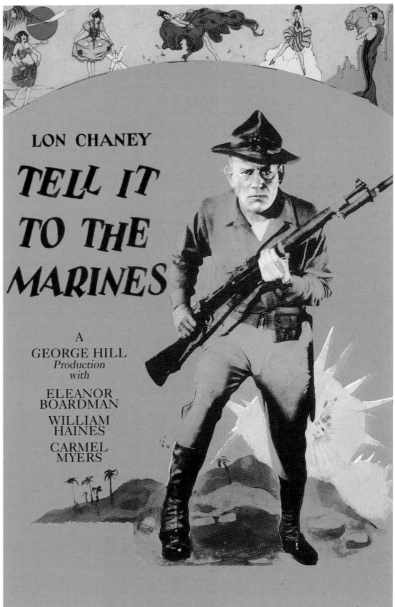

LON CHANEY

TELL IT TO THE MARINES

A
GEORGE HILL
Production
with

ELEANOR
BOARDMAN

WILLIAM
HAINES

CARMEL
MYERS

US Destroyer Base San Diego was established on February 23, 1922, south of the modern-day San Diego-Coronado Bridge. The designation culminated a campaign by Adm. Roger Welles, known as San Diego's "Navy Mayor," and others. It marked a major Navy commitment to San Diego. The *Sausalito News* reported that dredging an anchorage 1,800 feet wide at a depth of 17 feet would accommodate 120 ships. Within months, more than 75 decommissioned postwar destroyers were moored at the base. The base grew rapidly when torpedo and radio schools were established. During the Depression, Capt. Chester W. Nimitz campaigned for federal non-Navy funds to offset defense-spending cuts. The Navy received $2 million for needed dredging from the Public Works Administration. By 1937, the base had added two additional tracts of land. The cost of its 29 buildings and associated improvements amounted to more than $3.2 million. (US Navy.)

Advances in military aviation design came quickly in the 1920s and 1930s. Often, the newest generation of aircraft was sent to Rockwell Field for capability verification in the skies over San Diego. One example was the Curtiss B-2 Condor. Only 13 were built and all were too large for most aircraft hangars at that time. Most were stationed at North Island when the Army expanded its bombardment capability at Rockwell Field. The Condor carried up to 2,500 pounds of bombs, six machine guns, and a crew of five and at a cruising speed of 132 miles per hour had a range of 805 miles. Introduced in 1927, it was retired only seven years later when continued advances in aircraft design overtook the fabric-covered Condor. However, some of the twin-engine design elements of the Condor were not too far removed from the twin-engine, straight-wing aircraft that would be flying over San Diego a few years later. (San Diego Air & Space Museum.)

The Marine Corps Recruit Depot evolved markedly in the era between the two world wars. When first established, broad tidal flats separated the complex from San Diego Bay, as this aerial shows. In 1933, the Marines were reorganized to become the Fleet Marine Force, whose primary mission was to seize bases for naval operations in war. Since the bulk of the Pacific Fleet was in San Diego, Fleet Marine Force's headquarters was moved to San Diego. MCRD underwent significant expansion in the late 1930s as war loomed. Storehouses, barracks, mess facilities, a post office, parade ground, medical facilities, roads, railroads, and hundreds of 16-man huts were constructed. By 1939, the base, even with ancillary facilities in La Jolla and Kearny Mesa for firearms training, had become overcrowded.

By 1930, one-third of the Navy's submarine tonnage was operating out of the San Diego destroyer base, after the Navy transferred Submarine Division 20 to San Diego three years earlier. Various divisions of submarines alternated transfers between San Diego, the Panama Canal, and Pearl Harbor. They were often part of division, fleet, and Navy-Army exercises throughout the 1930s. Oftentimes, submarine tenders, such as the USS *Holland*, steamed into San Diego Bay, where as many as a dozen submarines tied up alongside. Acting as a floating ship repair facility, crews of the *Holland* and other tenders worked on the submarines in the bay. On other occasions, submarines tied up two and three deep at Navy Pier to take on supplies and crews.

In the late 1920s, it became clear that aircraft carriers would play an important role with the fleet. On the West Coast, the Navy's larger ships (including aircraft carriers) were based at San Pedro and not San Diego, largely due to San Diego Bay's relatively shallow waters. Once again, the San Diego Chamber of Commerce led a campaign for both federal funds and voter approval of bonds to finance more dredging. Federal funds were allocated and local voters passed a bond issue in 1928 to continue dredging San Diego Bay, in part to widen its narrow entrance. This photograph shows the narrow entrance to the bay, with North Island on the right and downtown San Diego in the distance. Widening and deepening the bay was necessary for large Navy ships.

The use of dirigibles by the Navy proved hazardous. On May 11, 1932, the USS *Akron* attempted to dock at Camp Kearny for refueling. On the fourth attempt, three members of the inexperienced ground crew were carried by mooring lines into the air. Within minutes, two fell to their deaths. The third, Bud Cowart, seen at right, held on for more than an hour before he was retrieved by the Akron's flight crew. News reports estimated Cowart rose up to 2,000 feet before he was rescued; the entire tragedy was captured by newsreel crews. The Navy ultimately decided to move much of its dirigible operations out of San Diego. (Both, National Archives.)

In the early 1930s, Rockwell Field remained relatively isolated, connected to Coronado only by a narrow causeway. Multiple landing strips and taxiways were often sprayed with oil to reduce dust from near constant use. Rockwell Field also became a battleground between the Navy and Army. Army chief of staff Gen. Douglas MacArthur wanted the Army to stay at Rockwell. That meant North Island's facilities would become more crowded as the Navy developed its nearly adjacent aircraft carrier facilities. But in 1935, Pres. Franklin Delano Roosevelt abolished joint use of four airfields and ordered Army operations at Rockwell Field to be transferred to March Field, about 100 miles away near Riverside. Naval Air Station North Island remains a principal Navy base nearly 80 years later.

San Diego's role in naval aviation aircraft carrier operations became more fully recognized when in the early 1930s nearly all aircraft assigned to the Navy's Pacific Fleet were transferred to North Island. That amounted to more than 350 fighters, bombers, and seaplanes. Formations such as these Grumman F2Fs operating between maintenance facilities on North Island and the USS *Saratoga* or USS *Lexington* off the coast became a common sight. By the dawn of World War II, naval aviation had developed such a prominent role that the Navy reorganized all of its aircraft into a single naval air forces command in 1942. Its headquarters was moved to San Diego in 1949, where it remains today. (Maritime Museum of San Diego.)

Shortly after leaving the Army in 1922, aviation pioneer T. Claude Ryan established the Ryan Flying School at Lindbergh Field. By 1925, he was operating the nation's first year-round, regularly scheduled commercial airline from San Diego to Los Angeles. Ten years later, another Ryan company, Ryan Aeronautical Company, built aircraft.

In the 1930s, Ryan developed the first monoplane trainer for the military. This version, the PT-22, became the standard for trainers. More than 1,000 PT-22s were purchased by the military by 1942; they were also used by several Allies during the war. By 1944, the innovative Ryan had developed the first jet fighter for the Navy.

Consolidated Aircraft Company became a major local source of jobs. More than 4,000 workers already were producing PBY Catalinas when the Navy placed an order for 200 new PBY aircraft in 1939, the largest Navy order since World War I. Consolidated took over former tuna canneries to add production capacity as pilots tested new designs over San Diego. (Maritime Museum of San Diego.)

San Diego's other great military aviation pioneer was Reuben H. Fleet. In 1935, he moved his Consolidated Aircraft Company from Buffalo to San Diego and opened a 275,000-square-foot factory at the edge of Lindbergh Field. He soon began building long-range flying boats (the most famous were PBY Catalinas) for the Navy.

The rudimentary Lindbergh Field in the early 1930s became a hub of military-related activity. For a time, the US Coast Guard borrowed part of a hangar for its fledgling operation that focused on countering smuggling across the Mexican border. On December 11, 1935, San Diego voters again ratified land donation for the military, this time 23 acres of tideland for a Coast Guard air station across the street from Lindbergh Field. Dredging created a square extension of new land out into the bay. When completed, a hangar, wooden seaplane ramp, mess hall, barracks, shops, and officers' quarters were constructed with Public Works Administration funds and became the Coast Guard's first air station in California.

By the time the Navy's pioneering aircraft carrier, the USS *Langley*, pulled up alongside Navy Pier, shown here in 1939, it had undergone several changes since its commissioning in 1922 and arrival on the West Coast two years later. It had developed early aircraft carrier flight operations in San Diego waters, including the first night landing. Replaced by larger and faster carriers, the *Langley* became a seaplane tender in 1936. Shortly after this photograph was taken, the *Langley* was sent to the Far East and, in 1942, was sunk after a Japanese aerial attack. By this time, Navy Pier in downtown San Diego had become a hub of the Navy's regional supply operation.

In 1939, Consolidated Aircraft expanded production by winning a contract for the B-24 Liberator bomber. Designed to outperform the B-17, early models cruised at 290 miles per hour for up to 1,700 miles and had a bomb capacity of 8,000 pounds. Ultimately, more than 18,000 Liberators were produced during the war, more than any other type of aircraft. They proved adaptable to bomber, transport, and patrol missions. About half were manufactured by Consolidated, and they became a common sight over San Diego. To meet wartime demand, Consolidated's San Diego facility was tripled in size and additional plants were built in the Midwest. (Both, US Air Force.)

This Consolidated "mystery plane" captured national attention when it was delivered to the military in 1939. With a 104-foot wingspan, it was widely speculated to be a new long-range bomber. By now, Consolidated had spawned an explosion of San Diego employment growth as the city's population surged to 192,500 by 1939. A shortage of housing developed as tent cities sprang up in Mission Valley. Water supply became an issue as annual industrial input climbed to $90 million. Four in 10 local workers were employed by the government or in aircraft manufacturing. Yet, for as large as aircraft manufacturing had become, the Army/Navy payroll was triple that of local manufacturing. The Navy was spending $40 million a year maintaining and operating its San Diego bases as war approached. (National Archives.)

For most of the 1930s, the USS *Saratoga* typified San Diego–based aircraft carriers that routinely transited San Diego Bay on their way to extended deployments or flight exercises a few miles off Coronado. On June 13, 1939, the *Saratoga* and the supply ship USS *Kanawha* completed an extended at-sea refueling exercise off San Diego, ushering in a new concept called underway replenishment; the Navy now knew that its carriers' range could be extended by months and thousands of miles.

OPPOSITE: Ryan Aeronautical Company operated three pilot schools in World War II, training more than 14,000 cadet pilots in San Diego. In addition, it diversified its aircraft design capability. In 1943, it produced the first piston-jet, mix-powered fighter, called the FR Fireball. Carrier suitability testing took place off San Diego in 1945. Ryan was one of eight Southern California wartime aircraft manufacturers that agreed to pool resources and inventions to help meet the overwhelming demand for aircraft.

BLITZ-BOOM
1940–1945

San Diego's long courtship of the military created a union in World War II that continues to define the region today. The confluence of world events, San Diego's strategic location, and unabated, voter-endorsed land giveaways made a permanent partnership unavoidable.

Shipbuilding and repair, as well as aircraft manufacturing, became the cornerstones of wartime military expansion. San Diego's rank as an industrial center climbed from 79th in 1939 to 28th only four years later. That spawned an unprecedented migration to San Diego of far-reaching ramifications. Previously isolated San Diego became home to hundreds of thousands of newcomers from every region in the country.

As a consequence, San Diego nearly became a federal city during the war. More than 20 military bases and housing projects overwhelmed local government. So the federal government stepped in with massive housing projects, capital, and city infrastructure investments that exceeded $100 million.

Military needs defined most city improvements. Many of the region's main transportation arteries today were developed with federal funds to meet military wartime needs. When water supplies proved insufficient, the Navy sought $7.7 million to build a new pipeline. Sagebrush-covered hills were flattened to accommodate thousands of hastily built homes, all with federal money. The result was what the *Saturday Evening Post* characterized as a "Blitz-Boom."

As World War II approached, the Navy's air station at North Island became a hotbed of naval aviation activity. The Pacific Fleet's aircraft needed to be maintained ashore, and new pilots had to be qualified for potential combat duty. In early 1940s, test pilot John "Jimmy" Thach, shown here, commanded the VF-3 squadron at North Island. He began working with another aviator, Edward O'Hare, on combat tactics. Thach first used match sticks on his dining room table to simulate aircraft formations and design new air combat techniques for American aviators. Their aircraft were inferior to Japanese aircraft at the time. Then Thach and O'Hare tested the new combat tactics in the air over North Island. The first real test came at the Battle of Midway in 1942. Thach shot down three aircraft on one mission, and the technique came to be known as the "Thach Weave." O'Hare later earned a Medal of Honor. (San Diego Air & Space Museum.)

On the eve of war, the military needed vast expanses near San Diego for training purposes. More than 26,000 acres 11 miles northeast of downtown (present-day Scripps Ranch area) became Camp Elliott on April 8, 1941. The Marines first trained there, and then the Navy began bussing its recruits there from the naval training center in 1944 for firearms training.

The Navy spent $1.5 million in 1942 to build barracks such as these, post exchanges, theaters, and even swimming pools at Camp Elliott. Aside from the main camp and Camp Linda Vista, the Green Farm Camp was used to train Marine scouts and snipers. The Jacques Farm Camp served as a tank school.

When Pres. Franklin Delano Roosevelt declared a national emergency in 1941, the Marines needed a large West Coast training facility for ranks that were expected to increase by the thousands. It had to be a vast tract of land suitable for simulated amphibious landings as well as for inland battle training. The Navy bought a 132,000-acre Mexican land grant ranch north of San Diego called Rancho Santa Margarita y Las Flores. In September 1942, President Roosevelt dedicated Camp Pendleton. It featured nearly 20 miles of undeveloped shoreline at the foot of the new 200-square-mile military reservation. Five major canyons, rolling hills, and mountain ranges were so rugged that security was possible only on horseback. The 9th Marine Regiment took occupancy of the new facility in 1942 when it completed a four-day, 40-mile march from Camp Elliott to Camp Pendleton. (US Marine Corps.)

Camp Pendleton's expanse enabled the Marines to conduct large maneuvers. Its flat beaches enabled practice assaults in Higgins boats. Amphibious training was critical to the estimated 60 percent of recruits who had never seen the ocean. In spite of Camp Pendleton's size, some recruits were sent 100 miles east into California's desert to train for North African warfare. Others were sent to Bing Crosby's Del Mar Track about 20 miles away for intense physical conditioning. Camp Pendleton also drew the attention of Hollywood. Here, a scene from John Wayne's *Guadalcanal* is filmed on the base. After the war, Pendleton remained a central training facility and a major economic driver in the region.

In 1940, the city deeded nearly 1,300 acres just north of La Jolla to the Army for a training facility. At first, Camp Callan's mission was to train long-range artillery crews to defend against a Japanese invasion. But by 1942, that mission shifted to antiaircraft training. In 1944, the camp's mission shifted to amphibious assaults. Camp Callan was a small city in the Torrey Pines area, replete with barracks, a 910-bed hospital, five post exchanges, five chapels, three theaters, and even a landfill. By the end of the war, 292,000 officers and men had fired an estimated two million rounds of ammunition at Camp Callan.

World War II was an economic boom for small business, nearly all of it in the service or manufacturing sectors that supplied military needs. For example, Howard "Skippy" Smith established a parachute manufacturing facility on Eighth Avenue in 1942. It was one of the few fully integrated businesses at the time. The flamboyant Smith had been a former air show stuntman. Before it closed in 1944, Pacific Parachute Company had produced thousands of parachutes as a subcontractor for another San Diego business that also was on Eighth Avenue: Standard Parachute Company. Standard Parachute supplied an estimated 150,000 parachutes to the military.

16271   10-6-43
CONSOLIDATED   VULTEE
EAST ADDITION TO BLDG.
NO. 7
CONTRACT NO. A-2

Consolidated Aircraft led an economic boom never before seen in San Diego. Total industrial output increased from $161 million in 1941 to $800 million the following year. As aircraft orders mounted, Consolidated Aircraft recruited thousands of employees, took over abandoned buildings, and protected itself against enemy attack. Miles of camouflage netting, such as seen in this image, was strung over the Consolidated complex, Pacific Highway, and Lindbergh Field. Some were painted to resemble innocuous city streets and innocent rooftops while portions were covered with painted chicken feathers. Regrettably, the feathers carried fleas and lice, requiring many employees to be deloused. The camouflage was part of a citywide defense plan that included air raids and blackouts until 1943, when fears of an enemy invasion faded.

In less than two years, women working at Consolidated Aircraft skyrocketed from 40 to 16,000. Recruiters mounted a local and national campaign to draw women out of the house and to San Diego with advertisements that read, "Should Your Wife Take a War Job? 5,000 Women Needed Now in Local War Plants! San Diego is a Vital War Production Area!" The San Diego Chamber of Commerce and leading merchants joined in a promotion called "Women in War Week." Marston's, Montgomery Ward, Sears, and others displayed storefront posters extolling women who worked. During the peak production months, 40 percent of Consolidated's 60,000 employees were women.

The San Diego workforce and city infrastructure was wholly unprepared for the tremendous numbers of new workers needed in World War II. Job recruitment programs targeted homemakers, extolling the virtues of jobs ranging from parachute makers to aircraft machinists. However, the resulting flood of 1,400 people arriving weekly in San Diego produced critical shortages of housing and schools.

While increasing numbers of mothers responded to help wanted articles, in September 1941, local schools saw 7,500 more students than had been in class three months earlier. Local school districts needed $4 million for new school construction. Although some residents resented the migrant defense workers they call "Aviation Oakies," the city council encouraged residents to rent rooms and garages to them as temporary housing.

San Diego government leaders scrambled to provide the city services, such as an expanded transportation system and utilities, to meet the needs of the tremendous influx of defense workers from across America. That expansion also included an expanded municipal workforce, such as bus drivers. San Diego became nearly bankrupt as the federal government took control of nearly 26 percent of all land in the city for a variety of military purposes, stripping $150 million of assessed valuation from the tax rolls. New suburban house development projects broke ground, placing additional strains on public transportation. The federal government was responsible for much of it, such as the 1,200-acre Linda Vista Housing Project, the largest public housing initiative in the country. Three thousand houses were built in only 200 days for 13,000 residents. Small towns a few miles from downtown San Diego seemed to be appearing almost overnight.

Youngsters conducted scrap paper drives in the face of shortages and rationing. Street vendors reused movie posters to post their prices. Shoppers were asked to bring their own baskets or bags. Merchants prominently displayed posters from the War Production Board that read, "Save Your Cans (and) Pass the Ammunition" and "Save Waste Fats for Explosives! Take them To Your Meat Dealer." San Diegans, of course, also were asked to buy war bonds to support the war. Rallies were held at Horton Plaza with entertainment, sometimes featuring the Melomen, an all-African American Navy band. All seven war bond drives met their national goals, totaling $135 billion.

In 1940, San Diego voters approved the transfer of another 21 acres, by a four-to-one margin, to the Navy for its stately Balboa Park medical complex. However, that was not enough to meet patient and medical training demand. Soon, Balboa Park's exposition buildings were handed over to the Navy. The hospital expanded from 56 buildings with 1,424 beds in 1941 to 241 buildings with 10,455 beds in 1945.

Balboa Park's house of hospitality became the nurses' barracks. Elegant landscaping belied a cold, drafty building that was repainted "mist grey" on the inside. Some of the 600 nurses put newspapers between their blankets to keep warm and often awoke to raucous lions and parrots at the zoo next door. (US Navy.)

It seemed every facility and public space throughout San Diego was appropriated by the military during World War II. Nearly all of San Diego Bay, Point Loma, Balboa Park, and massive tracts of undeveloped land had come under federal control. Dozens of specialist schools were established throughout the city. Public facilities were used as well. Here, recruits practice their abandon ship skills by jumping off Crystal Pier in Pacific Beach in 1942. Oceanside Pier was used for such purposes as well. San Diego Bay became particularly congested. In 1944, an estimated 3,000 tons of ammunition was loaded onto 200 ships every month.

The seeds of San Diego's modern defense technology industry were planted in World War II. Much of it took place behind locked gates, guarded fences, or in remote regions of San Diego County. In 1940, the Navy Radio and Sound Laboratory began working with the Scripps Institution of Oceanography and University of California on several initiatives. Underwater sound research and sonar design was relocated from the noisy San Diego Bay to this reservoir in 1943. Sweetwater Reservoir's depth, remoteness (about 15 miles east of San Diego), and solitude made it an ideal field research station. This led to sound transducer technology to enhance submarine detection. Other joint projects included Scripps Institution research into surf condition predictions. That work supported the Allies' landings in North Africa and across the South Pacific.

A strange plane took off from North Island on September 20, 1942. It was this Japanese Zero that had been recovered from an Alaska island earlier in the year when its pilot crashed into a bog and was killed. After repairs, it took off from North Island and flew simulated dogfights against Wildcats, Corsairs, and other aircraft in the area. American pilots discovered design weaknesses in the generally superior Zero, giving them an important tactical advantage for the rest of the war. The heavily guarded initiative marked a classified coup for North Island in the war. The Zero was accidently destroyed in February 1945 when another aircraft taxied into it.

This barrage balloon in Balboa Park in 1942 was part of San Diego's early war defense network. The 307th Coast Artillery Barrage Balloon Battalion sent three units to San Diego in early 1942. Balloons were moored near the Consolidated Aircraft plant, Lindbergh Field, and Balboa Park as a deterrent against enemy aircraft attack. Connected by cables when aloft, protection of a two-mile-wide target area required 100 balloons at 5,000 feet. Crews were on 24-hour alert for about a year. Elsewhere, antisubmarine nets were strung across the mouth of San Diego Bay and longer-range artillery was installed at Fort Rosecrans and elsewhere along the coast. The balloon battalion was disbanded in September 1943 when the threat of attack by the Japanese was deemed no longer credible.

San Diego's prosperous tuna fleet became the "Pork Chop Express" in World War II. Nearly 50 volunteer tuna clippers became supply ships, hauling mostly meat, vegetables, and sometimes holiday turkeys from Pearl Harbor to western Pacific islands. One group delivered badly needed fuel to a remote atoll only days before the pivotal Battle of Midway. Their large freezers, small crews, and long range made them perfect for 1,800-mile voyages. With a top speed of only about 10 knots, most sailed in small convoys and were equipped with machine guns and a few depth charges. Of the 49 clippers, 17 were lost to storms, enemy fire, and accidents. Survivors came home with knowledge of radar, advanced depth-finding sounders, and enhanced refrigeration systems.

San Diego boosters continually sought opportunities to promote the city to a national audience. Throughout the late 1930s and again in 1941, seen here, the city was represented in the annual New Year's Day Rose Parade. A year later, the USS *San Diego*, a light cruiser, was commissioned. The second Navy ship called the USS *San Diego*, it served with distinction and was the first major Allied ship to sail into Tokyo Bay in 1945 at the end of the war. With 18 battle stars, it was one of the most decorated ships in the war. (Maritime Museum of San Diego.)

Other than by attire, military personnel and San Diego residents almost became indistinguishable throughout the city. Tens of thousands of Marines and sailors rotated through the city every few weeks. Other than Belmont (amusement) Park, the beaches, and Balboa Park, nearly all entertainment was centered along Broadway. Peep shows, bars, shoeshine stands, military uniform tailors, movie theaters, souvenir photographers, and restaurants all beckoned. Here, the Bomber Café at 849 Fourth Avenue pays tribute to the B-24 bombers being manufactured only a few miles away. A turkey or spam sandwich cost 20¢ and a soft drink or coffee cost a nickel. Almost next door, Sam the Tailor offered tailored Navy uniforms and a "clean while you wait" service.

San Diegans joined the nation in celebrating peace in 1945. San Diego would never be the same, as many veterans stayed after discharge. The Navy began taking over houses recently abandoned by laid-off aircraft manufacturing employees as the city's population shrunk markedly. By the end of 1945, the Navy was consuming 17.7 billion gallons of municipal water, 40 percent of the city's supply. For its part, San Diego welcomed the military's continued presence. Plans were formulated for a Veterans Memorial District spanning six blocks downtown that would include a convention auditorium. It became a controversial measure due to the cost. Voters rejected a tax measure to finance the $3 million project in November 1945. A Navy chapel near Balboa Park later was transformed into a Veterans Museum in 1989.

Although tens of thousands of San Diegans attended this 1945 Navy Day parade along Broadway, San Diego's economy was already teetering on a recession by the end of the war. The city's population had dropped from 385,000 in 1943 to 286,000 in 1944 as military manufacturing operations were scaled back. When Camp Callan was closed and the lease with the city expired, a cash-strapped city council bought the lumber from 500 dismantled buildings. The city resold it for new housing for veterans at a $250,000 profit. The national recession that followed World War II was less severe in San Diego, however, largely due to the city's profound reliance on military spending. (Maritime Museum of San Diego.)

OPPOSITE: Some of the nation's foremost radar and radio experts worked at the Navy Radio and Sound Laboratory atop Point Loma during the war. Accomplishments included the first USN radar set, improved aircraft radio reception, worldwide navigation beacons, and submarine decoy beacons. Classified postwar research requirements resulted in significant expansion of the facility, including the conversion of former Fort Rosecrans batteries into laboratories. The breadth of its work led to it being renamed the Navy Electronics Laboratory (NEL) in 1945. (Space & Naval Warfare Systems Center Pacific.)

Six

COLD WAR WARRIOR
1946–1959

As the growth bubble that had powered San Diego through World War II collapsed at the end of the war, a new opportunity presented itself: the Cold War. It came at a time of what one historian called a "Revolution in Naval Affairs." Supersonic jets, guided missiles, nuclear weapons, and nuclear-powered ships represented the future.

San Diego had become a de facto federal city during the war where the government funded new suburbs, roads, dredging, and land creation along San Diego Bay and even an expanded water system. Now, as the postwar population plummeted and aircraft production orders were cancelled, San Diego stood at a crossroad.

It would have to reinvent its fundamental relationship with the military. A city historically in quest of more military bases would have to transform itself into an educated metropolitan region worthy of significant defense science research-and-development investment. With 21 military installations in the area, it would have to become a Cold War warrior in what its congressman called the "most militaristic district in the country."

In World War II, San Diego solidified itself as a strategic Navy base of operations. That continued after the war as the Navy's San Diego payroll of $117 million in 1950 dwarfed a $30 million Navy payroll in 1939. In 1949, approximately 450 Navy ships comprised a mothball fleet in San Diego Bay. More than 2,000 civilian employees maintained them until many were reactivated at the outset of the Korean War. During the Korean War, the number of active-duty Navy ships in San Diego increased from 42 in 1950 to 146 in 1952. Meanwhile, as the active-duty fleet expanded, ships routinely tied up to Broadway Pier and Navy Pier, shown here, to take on men or supplies.

Some of the World War II housing built for aircraft manufacturing employees, such as this tract, became available for the postwar Navy. City leaders pursued economic diversification via military R&D, federal investment in military bases, development of higher education facilities, and heightened tourism. Postwar San Diego never backed away from its unabashed support for the military while looking to build a broader economic base.

Following the war, part of the San Diego Naval Hospital's focus shifted to long-term rehabilitation for veterans, like these amputees. Many veterans who had been wounded stayed in San Diego after their discharge, as advances in prosthetics were developed to meet the high demand. Similarly, at one point in the early 1950s, two-thirds of the local physicians had arrived in San Diego while serving in the Navy. (US Navy.)

San Diego's water supply today is a direct descendant of the military's presence here. In 1947, the city faced chronic water shortages in light of increased military water consumption. With the city unable to pay for a new pipeline, the federal government took control by funding construction of the pipeline in return for preferential water supply and rate treatment. The federal government also selected an aqueduct route (tapping into a Los Angeles–area pipeline) most advantageous to its bases. Upon completion, the pipeline was almost immediately at capacity, and a second barrel was built with federal funds made available in 1952. About five years later, that too was at capacity. Regardless, the military had a more reliable water supply, and San Diego was forced into significant dependence on buying most of its water from the Metropolitan Water District of Southern California.

Part of World War II's legacy in San Diego was the city's entrenched position as the West Coast's foremost military training, supply, and operations center. As a result, San Diego found itself "on the front line" when the Korean War broke out. In August 1950 alone, 16,000 Marines shipped out of San Diego. Here, nine Navy ships are moored almost side by side in downtown San Diego. Dozens of war ships departed San Diego for combat duty in the Far East as Navy personnel in San Diego increased by 500 percent in the early 1950s. Meanwhile, at North Island's Ream Field and Brown Field, 11 aircraft squadrons were activated and established wartime operations. (Maritime Museum of San Diego.)

Marine recruit training was not confined to MCRD near Lindbergh Field. Nearby commercial development required weapons training to take place about 11 miles north on an undeveloped tract near La Jolla. Beginning in 1917, thousands of young Marines honed their skills at the Weapons Training & Small Arms Facility. At its peak in World War II, 9,000 recruits learned how to handle firearms there every three weeks. A few years after the Korean War, San Diego officials launched a campaign to make the Camp Matthews acreage available for a new University of California campus. After initial resistance by the Marines, a new weapons training facility was built on a former wheat field at Camp Pendleton. Camp Matthews closed in 1964 and today is part of the University of California, San Diego campus.

Camp Pendleton reflected the level of military activity in the San Diego region in the years following World War II. After becoming one of the largest military reservations in the world, by the end of 1946 military personnel had declined from a wartime high of 86,000 to 4,500 active-duty personnel. The 40,000 sheep grazing on Camp Pendleton far outnumbered them until the outbreak of the Korean War. Within four days of the war's outbreak, lead Marine elements from Camp Pendleton departed from San Diego to reinforce troops at Pusan. Soon, a Korean-style town was built at Camp Pendleton to train a reconstituted 1st Marine Division. An estimated 200,000 Marines had trained at camp Pendleton for combat in Korea by June 1953. Camp Pendleton remains a primary training base for Marines bound for overseas duty.

Postwar military radio communications becoming more complex prompted interference problems for military ships and aircraft. Beginning in 1947, NEL engineers experimented on this Antenna Model Range to develop antennas that could receive simultaneous communications on different wavelengths. They positioned 1/48-scale ships on a 22-foot, brass-coated roundtable underneath a zenith arch antenna that could be shifted to different angles. Engineers tested various ship antenna configurations to eliminate self-interference and to reduce the number of antennas needed on ships. Four years later, the USS *Mount McKinley* was commissioned with one-third the number of antennas found on its predecessors with no loss in performance. (Both, Space & Naval Warfare Systems Center Pacific.)

Just as San Diego transformed itself into a Navy town in the early 1900s, it faced another transformation in the late 1940s. Civic leaders realized San Diego could no longer afford to be a two-industry city after the near total collapse of the aircraft manufacturing industry following VJ Day. The dawn of the jet age, guide missiles, and the aerospace industry required a shift from assembly-line aircraft production to sophisticated defense R&D. That reality was driven home in 1954 when General Dynamics bought Consolidated Aircraft (which had become known as Convair after a previous merger). Rocket development offered new employment opportunities. National employment campaigns were conducted, extolling the virtues of living in San Diego, employer housing assistance, and opportunities for manufacturing technicians as well as in emerging military science professions. (San Diego Air & Space Museum.)

It took 10 years of Project Atlas development before Convair unveiled its signature Atlas rocket in 1957. That rocket's flight lasted only 24 seconds before losing thrust and self-destructing. Regardless, the United States now stood ready to explore and arm the edge of space. Originally developed for the Air Force as an intercontinental ballistic missile, the SM-65 Atlas could deliver a nuclear warhead and had a range of 5,000 miles. Its innovative, lightweight design featured a surprisingly thin shell that had structural integrity provided by internal high-pressure fuel tanks. An Atlas rocket launched the world's first communication satellite in 1958 and became an operational ICBM in 1959. Ultimately, the San Diego–built Atlas powered 10 of the Mercury space missions. In addition, the Atlas-Centaur variant launched numerous geosynchronous communication satellites and various space probes. Here, a technician inspects the inner housing of a rocket assembly. (San Diego Air & Space Museum.)

In the 1950s, San Diego stood at the center of warfare potentially waged across the ionosphere or from under the polar ice caps. The NEL's Dr. Waldo Lyon developed a fathometer that enabled submarines to navigate under the ice cap, once submarines no longer needed to surface for ventilation. Shown here on a USS *Nautilus* voyage in 1957, his equipment enabled *Nautilus* to travel more than 1,000 miles under the ice on this and another voyage the following year. About the same time, the USS *Skate* became the first submarine to surface at the North Pole. This defense-inspired research revealed the polar ice cap was about five times thicker than previously thought and that ice keels could extend up to 150 feet down from the surface. (Both, Space & Naval Warfare Systems Center Pacific.)

San Diego's rich heritage as a nexus of seaplane development continued in the early 1950s with the Sea Dart. It could be deployed rapidly and as a seaplane did not require expensive runways and military bases. Performance was disappointing, and one Sea Dart broke up during a November 4, 1954, demonstration flight over San Diego, killing the test pilot. The Sea Dart project was scrapped in 1956, and the seaplane base idea was dropped the following year. The Sea Dart remains the only seaplane to break the sound barrier.

Like several postwar aircraft, the F-102 Delta Dagger built by Convair included design elements taken from German delta-wing research and development during World War II. It was designed to be an all-weather supersonic interceptor of long-range Soviet bombers during the Cold War. As the first operational delta-wing aircraft in the world, it was also the first all-missile fighter. After its maiden flight in 1954 that demonstrated a climbing ability of 50,000 feet in four minutes, nearly 1,000 Delta Daggers ultimately were accepted into the US military inventory. More than 100 were also built as twin-seat trainers. Some of those provided demonstration flights for celebrities such as Arthur Godfrey and Rock Hudson as part of Convair's high-profile public relations campaign. Delta Daggers remained operational for about 20 years before they were replaced by the F-106 Delta Dart that could fly at Mach 2.

Throughout the 1950s, Convair developed dozens of futuristic military aircraft designs. One was the XFY Pogo, an aircraft that could take off and land vertically. Both the Air Force and Navy expressed interest in the concept at various points in its development, leading to the first test flight in 1954. The prototype was tested at Brown Field, southeast of San Diego. The Navy envisioned the Pogo on its cargo ships and possibly submarines. The Vertijet was designed for 20-millimeter cannon or 48 unguided rockets on wingtip pods. Extremely complex components and an untested power plant made landing particularly dangerous. Ultimately, the project was abandoned after three were built and one was flown. Convair donated the prototype to the Smithsonian in 1960. (San Diego Air & Space Museum.)

San Diego needed another William Kettner in Congress in the 1950s as the city sought to reinvent itself during the Cold War. Rep. Bob Wilson (at left) answered that need, serving for 14 terms beginning in 1952. As a member of the House Armed Services Committee, the former advertising executive secured funding to dredge two million cubic yards to accommodate new super carriers, such as the USS *Kitty Hawk*. The dredged sediment was used to create Harbor Island. He also played a key role in the transferring of 280 acres at Fort Rosecrans to the Navy, in part for a permanent nuclear submarine base at Ballast Point. Here, the USS *Grayback* returns to San Diego Bay with an early guided missile.

Defense research in the late 1950s required a more detailed understanding of the ocean. This oceanographic research tower was built in 1959 by the Navy Electronics Laboratory a few miles off Mission Beach. It provided laboratory-like conditions with 150 external sensors to study water motion, underwater acoustics, electromagnetic propagation, marine chemistry, marine biology, marine geology, and to evaluate the latest Navy subsurface technology. All were critical to advances in subsurface navigation and transmission. In 1965, a proposal to build a second tower in deeper water stated the cost would be less than 10 percent the expense of a new seagoing oceanographic research vessel and ongoing maintenance also would be significantly less expensive. It was never built, however. The original tower was destroyed by a January 1988 storm, two years after the Scripps Institution of Oceanography had taken custody. (Space & Naval Warfare Systems Center Pacific.)

Throughout the 1950s, Convair engineers developed new and sometimes radical aircraft concepts in response to changing military requirements. The Tradewind began as an antisubmarine warfare aircraft, then was designed to lay mines, and finally became a transport for up to 90 troops. Convair engineers also developed nuclear-powered aircraft concepts as part of the Department of Defense's Aircraft Nuclear Propulsion Program. Government contracts of more than $70 million enabled Convair engineers to develop airframe designs in conjunction with nuclear power plant designs devised by General Electric. More than $1 billion was spent on this program before it was discontinued in 1961.

Perhaps no one was more responsible for the partnership between the Navy, Scripps, and the University of California that defined San Diego's reinvention as a defense scientific community than Dr. Roger Revelle. He pioneered radar and sonar research at the Navy Radio and Sound Laboratory during World War II and worked on the University of California's Department of War Research projects that were managed by the Scripps Institution of Oceanography. At the Office of Research and Inventions, he awarded military research contracts to Scripps. Revelle later became director of Scripps and led the campaign to establish the University of California, San Diego.

OPPOSITE: Modern-day combat ranges from the deployment of stealth aircraft to unmanned aerial vehicles. UAVs trace their roots back to World War I's primitive attempts at radio-controlled aircraft. By the mid-1930s, the concept of radio-controlled "drone" aircraft had been developed and the Navy concentrated its drone research programs in San Diego. Stealth technology became critical due to World War II's radar technology advances, many of them made by Navy researchers in San Diego. In the 70 years since World War II, several San Diego defense contractors have become leaders in UAV and stealth technology. (Department of Defense.)

SEVEN

NETWORK-CENTRIC WARFARE
1960–1990

San Diego remained a national center of military research and development as the Cold War intensified in the 1960s and 1970s. With rapid technology advancements, the city entered into a new period that has been termed the defense science-and-technology era.

Clusters of entrepreneurial defense contractors emerged, often due to new military science needs, including advanced combat direction systems, satellite laser communication, weapons systems, mine neutralization, unmanned sea and air vehicles, and light detection and ranging, among others. By 1990, they became major employers and centers of civic and national leadership.

Meanwhile, national defense continued to evolve toward the concept of network-centric warfare, which was premised on processing unimaginable amounts of data almost instantaneously to provide dominant battlefield awareness. On the eve of Operation Desert Storm, both the military and its San Diego defense contractors were beginning to rewrite the rules of war based on information superiority instead of military mass.

Once again, San Diego began an extended reinvention in response to world affairs and national priorities, not unlike the era when the Great White Fleet arrived in 1908. Indeed, the vision of those intrepid civic leaders more than a century ago had become reality. San Diego had grown from a presidio to a Pacific powerhouse, poised to play a vital role in America's 21st-century defense.

The Navy established the Deep Submergence Systems Project (DSSP) following the loss of the USS *Thresher*. It conducted SeaLab II experiments off La Jolla. Three groups of 10-man teams lived and worked for 15 days at a time at a depth of more than 200 feet. The undersea habitat measured 50 by 12 feet, divided into an entry, laboratory, galley, and living spaces. Research was conducted on human performance, oceanography, and Navy salvage potential. Other work included undersea tool testing, salvaging a Navy fighter jet, mining ore samples, and studying ocean-floor geology. On some missions, trained dolphins delivered messages, the mail, and tools to SeaLab II.

Submarine Development Squadron Five was first organized in 1967 in San Diego. The research submarine USS *Dolphin* was commissioned the following year and assigned to the Naval Undersea Warfare Center on Point Loma. It could transport 12 tons of testing equipment and conducted the first launch of a Mobile Submarine Simulator System (MOSS) device. MOSS was a self-propelled decoy that was launched through torpedo tubes and was a descendant of the first submarine decoy beacons developed in San Diego during World War II. Squadron Five continues to work with scientific and academic institutions on military-requirement initiatives, including escape, rescue, special warfare, and unmanned underwater vehicles. Decommissioned in 2007, the *Dolphin* is now open to the public at the Maritime Museum of San Diego. (Scott McGaugh.)

The success of its Mercury Atlas rocket program led General Dynamics to open a massive plant in Kearny Mesa, a neighborhood of San Diego, in 1958. That required a national hiring campaign and provided a significant boost to the local economy. On average, an Atlas rocket was launched every month throughout the 1960s. General Dynamics later lost its bid to produce Apollo mission rockets, prompting layoffs that reduced its local workforce from 40,000 to 15,000 employees. It diversified into cruise missile production before selling its Space Systems Division to Martin Marietta Corp. for $208.5 million in 1993. The bulk of local missile manufacturing was relocated to Arizona and Colorado.

In 1972, General Dynamics' Kearny Mesa plant began designing a cruise missile called the Tomahawk. Manufacturing began in 1976. The Tomahawk became a cornerstone of the military's arsenal for the next three decades. By 1991, Tomahawks were 19 feet long, weighed more than 3,000 pounds, and cost approximately $1.1 million each. In the Gulf War, about half of the 284 Tomahawks launched had been built in the Kearny Mesa plant. Armed with up to a 1,000-pound warhead, they achieved a 90 percent direct-hit success rate, according to Pentagon officials. More than 2,000 have been launched in combat. (San Diego Air & Space Museum.)

Military research in San Diego has directly impacted NASA's space programs. In 1969, the Naval Electronics Laboratory Center (NELC) provided satellite communication technology to NASA after atmospheric interference had interrupted Apollo 8 radio transmissions. NELC developed a portable receiver/transmitter unit in one month's time. It was installed aboard the USS *Guadalcanal* in time for Apollo 9. The unit became the primary command control circuit between Mission Control in Houston and the military in the recovery area. NELC engineers operated the equipment for that mission and the two Apollo flights that followed before NASA took control. A Navy post-mission report stated, "Apollo 9 communications in the Atlantic Command Area were a textbook example of an efficient error free operation." (Space & Naval Warfare Systems Center Pacific.)

In the late 1950s, the Navy was beset by multiple radio navigation systems. One system, Loran, required 57 stations worldwide but covered only about 10 percent of the earth's surface. Its navigational fix accuracy was plus or minus two miles. Building on the radio communications work that began in San Diego 50 years earlier, the Navy Electronics Laboratory developed a new system in the 1960s that utilized very low radio frequencies. This system, called Omega, proved to be more reliable regardless of weather and could reach submarines below the ocean's surface and under the polar ice cap. Aircraft and ships could develop a navigational fix to within one-half mile during the day and one mile at night. First operational in 1968, Omega was made available to all nations in 1990. (Space & Naval Warfare Systems Center Pacific.)

The Navy Electronics Laboratory established this Minitrack Station, Brown Field (southeast of San Diego), as part of the Navy's first network designed to track satellites, called Project Vanguard. This facility was the first American satellite station to track Russia's orbiting *Sputnik*. A few years later, it was transferred to another location in the Mojave Desert. (Space & Naval Warfare Systems Center Pacific.)

Four years later, the Navy Electronics Laboratory built the La Posta Astro-Geophysical Observatory in the Laguna Mountains at an elevation of 3,900 feet and about 65 miles east of San Diego. As part of NEL's microwave research, this facility conducted solar radio mapping and solar flare research. (Space & Naval Warfare Systems Center Pacific.)

The development of undersea vehicles required greater surface ship stability in high seas. Dr. Thomas Lang began developing a new ship concept in San Diego in 1968 at the Naval Undersea Warfare Center. The result was the SSP *Kaimalino*, the first high-performance SWATH (Small Waterplane-Area Twin-Hulled) oceangoing vessel. Potential applications included utility boat functions, oceanographic research, antisubmarine warfare, submersible tender, and helicopter transport. Several of the Navy's newly design undersea surveillance ships used this design to achieve exceptional stability in a small platform. (Department of Defense.)

In the 1960s, remotely operated vehicles (ROVs) significantly expanded the military's undersea presence and capability. The first cable-controlled Underwater Recovery Vehicle (CURV I) was developed in 1965. The following year, it recovered a thermonuclear hydrogen bomb off the coast of Palomares, Spain. This CURV III variant operated by Naval Ocean Systems Center, San Diego, became a Navy workhorse for more than 20 years. In 1973, CURV III rescued two Canadian crewmen trapped in the Pisces-III submersible off Ireland at a depth of 1,375 feet. Over the years, CURV III has participated in the recovery of various classified objects and conducted many secret missions during the Cold War. At least one variant of CURV could descend more than 20,000 feet. (Space & Naval Warfare Systems Center Pacific.)

Lesser-known San Diego R&D firms have contributed to military capability, including Ametek Straza. It developed sonar systems for nuclear-powered submarines and an ROV called Scorpio (Submersible Craft for Ocean Repair, Position, Inspection and Observation). Scorpio was connected to a surface ship by a flexible umbilical cable. In this image, Scorpio is tested in San Diego Bay. In addition to tethered ROVs, the Ocean Engineering Division at the Naval Undersea Center/Naval Ocean Systems Center began developing advanced, unmanned, and untethered ROV search capability in 1973. By the mid-1990s, this type of ROV had completed 114 launches and recoveries without mishap. Improvements in underwater acoustic communication were vital to conducting more sophisticated untethered ROV searches. (Jan Kocian.)

Military training in San Diego has included sea lions, whales, and dolphins for more than 40 years. The Marine Mammal Program was established in 1960, and the San Diego facility became operational in 1967. Sea lions' underwater directional hearing has enabled them to recover test missiles and other objects. In a project called Deep Ops, pilots and killer whales recovered objects at depths of more than 1,600 feet. Dolphins' biological sonar has been used to mark mines and detect enemy swimmers. The primary training facility today is at Point Loma, and training frequently takes place under the USS Midway Museum in San Diego Bay. (Both, Department of Defense.)

Naval Amphibious Base Coronado was an ideal training location when President Kennedy pledged $100 million for special operations warfare development in 1961. The first SEAL (Sea Air Land) team was commissioned on January 1, 1962, at NAB Coronado. Worldwide command of all Navy SEALs is located at Coronado as well. SEALs trace their roots to the scout raiding units in World War II and Navy Underwater Demolition Teams that helped clear beaches in advance of amphibious landings. Renowned for their toughness, dedication, and absolute secrecy, SEAL teams regularly can be seen transiting through San Diego Bay out into the Pacific on various training exercises. (Both, Department of Defense.)

The Navy took control of the former Marine Corps airfield north of Kearny Mesa in 1952. Naval Air Station Miramar gained international fame when the Top Gun school was established there on March 31, 1969. At the time, military aviators had received minimal air-to-air combat training in an era when many thought air-to-air missiles had rendered dogfighting obsolete. Early-war losses over Vietnam proved the opposite when kill ratios were 2 to 1 compared with 10 to 1 in World War II. Soon, Top Gun graduate pilots were dominating the sky, in part because they had been able to train with and against captured enemy MiG-17 and MiG-21 jets. As program graduates, they became training officers in various squadrons. In 1996, Top Gun relocated to Fallon, Nevada, a blow to the civic pride for many San Diegans. (Department of Defense.)

The fall of Saigon on April 30, 1975, spawned a major migration of Southeast Asians to the United States, many of them through Camp Pendleton. Pendleton was one of four processing centers and was comprised of nearly 1,000 tents and almost 150 Quonset huts to house refugees. Between April and November, 165 births took place across the four separate camps. Ultimately, more than 50,000 Vietnamese refugees passed through Camp Pendleton at a cost of approximately $17 million. Thousands were sponsored by San Diego families and organizations. According to the 2008 census estimate, 37,000 Vietnamese were living in San Diego County. (Both, Elisa Leonelli.)

General Atomics and other defense contractors in San Diego have been at the forefront of unmanned aerial vehicle (UAV) development, dating back to the late 1980s. At that time, San Diego–based General Atomics acquired the technology of a UAV pioneering company that had gone bankrupt. Only about five years later, its Predator RQ-11 was the first UAV deployed to the Balkans. In the 21st century, the increasingly sophisticated UAV military and civilian markets have become a multibillion dollar industry. Now one of its leaders, General Atomics was founded in 1955 in San Diego. A state-of-the-art research campus was built on Torrey Pines mesa after voters approved a land transfer in 1959. Today, a San Diego–built UAV can distinguish a milk carton on a picnic table from 65,000 feet and can fly for 40 hours over the battlefield.

Local military installations are no longer located on isolated expanses of sagebrush, cactus, and sumac in San Diego County. This aerial of NAS North Island illustrates how San Diego's military bases have become fully integrated with suburbs and communities that typically look on them with pride. The Marines' Camp Elliott is now Tierrasanta and Scripps Ranch. Camp Matthews on the Torrey Pines mesa now is part of the University of California, San Diego campus. By the start of the 21st century, the military was spending $10 billion in San Diego on contracts, military pay, and veterans' benefits. The 14 military bases in the area generated approximately 300,000 jobs, fulfilling the nearly 100-year-old vision of William Kettner, John Spreckels, and even Teddy Roosevelt.

The San Diego–based hospital ship USNS *Mercy* reflects the deep relationship between the Navy and the city's shipbuilders. San Diego's shipbuilding industry, largely founded on military needs once San Diego became a strategic Navy homeport, has ebbed and flowed through eras of war and peace. The National Steel and Shipbuilding Company (NASSCO) built the SS *Worth* oil tanker in 1976 and then retrofitted it into the world's most sophisticated Navy hospital ship, the *Mercy*, in 1984. The 1,000-bed floating hospital has completed numerous humanitarian missions around the world and has supported such military operations as Operation Desert Shield. More than 100,000 patients were treated, and more than 1,000 surgeries have been performed on *Mercy*'s humanitarian deployments to Southeast Asia. NASSCO continues to be a significant shipbuilder for the Navy. (Department of Defense.)

Much of San Diego's military-industrial complex is not readily accessible. Clusters of antennas are mounted atop the former World War II–era Consolidated aircraft manufacturing plant, still a landmark along Interstate 5 at the north edge of downtown San Diego. Today, those former plants are the heart of the Space & Naval Warfare Systems Command (SPAWAR). The western end of Point Loma remains a closely guarded military reservation. Point Loma facilities are now part of SPAWAR Systems Center Pacific. While San Diego remains a strategic homeport for both the Navy and Marines, San Diego's network-centric military research focuses on command, control, communications, computers, intelligence, surveillance, reconnaissance, lasers, and other classified initiatives. (Both, Space & Naval Warfare Systems Center Pacific.)

DISCOVER THOUSANDS OF LOCAL HISTORY BOOKS
FEATURING MILLIONS OF VINTAGE IMAGES

Arcadia Publishing, the leading local history publisher in the United States, is committed to making history accessible and meaningful through publishing books that celebrate and preserve the heritage of America's people and places.

Find more books like this at
www.arcadiapublishing.com

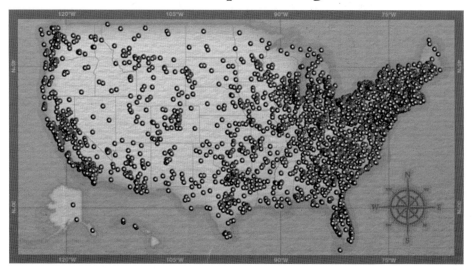

Search for your hometown history, your old stomping grounds, and even your favorite sports team.

Consistent with our mission to preserve history on a local level, this book was printed in South Carolina on American-made paper and manufactured entirely in the United States. Products carrying the accredited Forest Stewardship Council (FSC) label are printed on 100 percent FSC-certified paper.

MADE IN THE USA